This Book Belongs TO:

....

....

Jaime Harmon Rourke
Dr. Rhonda Y. Green Harmon
MAPLE AND BIRCH PUBLISHING, LLC

NUMBER CHART (1 – 100)

Fill in the missing numbers and count to 100. Enjoy this fun activity. Yes you can count!

1	2		4	5			8		10
11		13			16		18	19	
21	22		24		26	27		29	30
		33	34	35		37		39	
41	42			45		47	48		50
51		53	54	55	56			59	60
61	62	63		65		67			70
71		73	74		76		78		
	82	83		85	86	87	88		90
91			94	95		97		99	

COUNT BY 1s

Fill in the missing numbers.

1

1			4					9		

2

21				25					30	

3

40										50

4

26								34		36

5

34			37				41			

6

75					80					85

COUNT BY 2s

Fill in the missing numbers.

1

2			8			14	

2

40				48					58	

3

22					32			38		

4

64				72						84

5

12			18				26			

6

74						86				94

COUNT BY 5s

Fill in the missing numbers.

5			20
	30		
			60
	70		
			100

COUNT BY 10s

Fill in the missing numbers.

10		30		50
30				70
20			50	
50				
40				

DIRECTIONS: FILL IN THE MISSING NUMBERS IN EACH SET USING THE CUES PROVIDED.

1		3	4	
6		8		

2	4		8	10
		16	18	

5		15		25
	35	40		

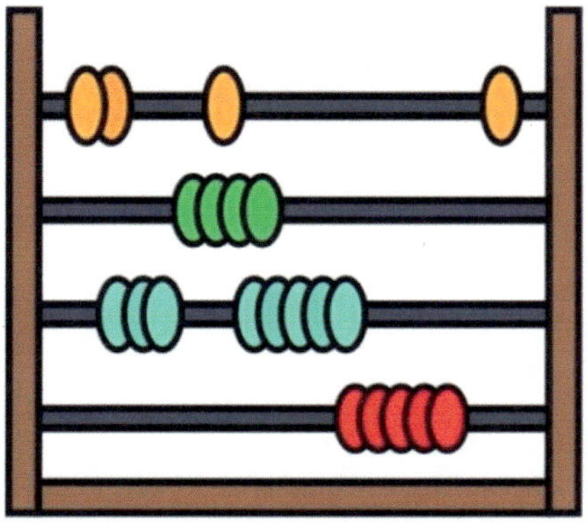

10	20		40	
60		80		

BEFORE AND AFTER

Write the numbers that come before and after the number in the middle.

5	6	7
	9	
	5	
	3	
	7	

MATCH THE INSECTS

CIRCLE THE PICTURE

Circle the cloud that has more raindrops and write the number in the box.

MATCH THE SHADOW

Draw a line to match the shadow of these school supply items.

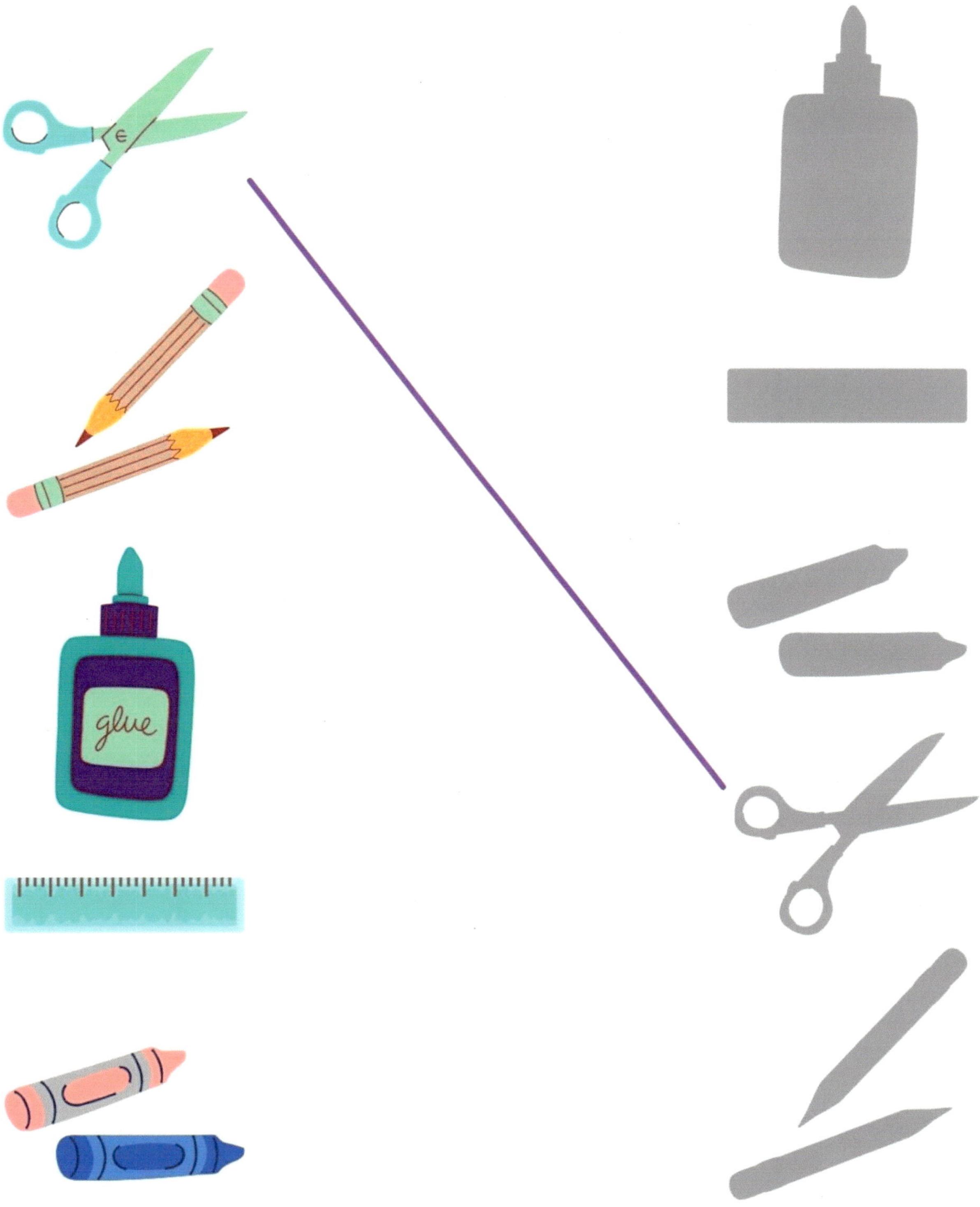

MATCH THE SHADOW

Draw a line to match the shadow of these items that might be found in a garage.

SIMPLE ADDITION

Add the items in each group and write the answer on the line.

3 + 2 = ______

1 + 1 = ______

2 + 2 = ______

1 + 2 = ______

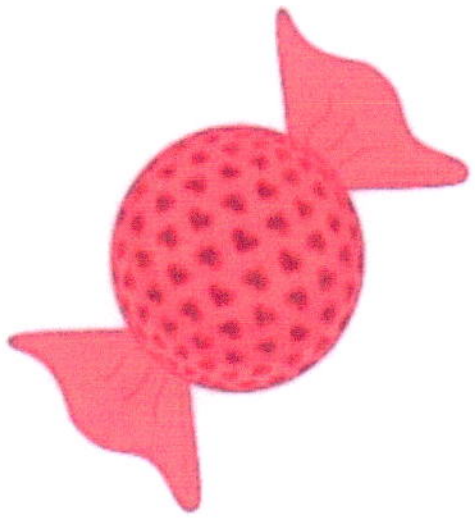

1 + 0 = ______

3 + 1 = ______

COUNT CANDLES

Solve the addition problems and write the correct answer on the line.

5 + 2 = ______	
3 + 5 = ______	
4 + 1 = ______	
3 + 1 = ______	
5 + 1 = ______	
1 + 2 = ______	

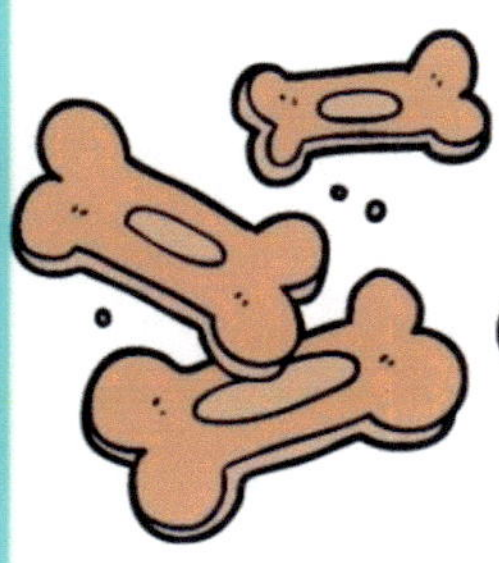

HOW MANY DOG BONES?

Count the bones carefully and put the correct number in the circle.

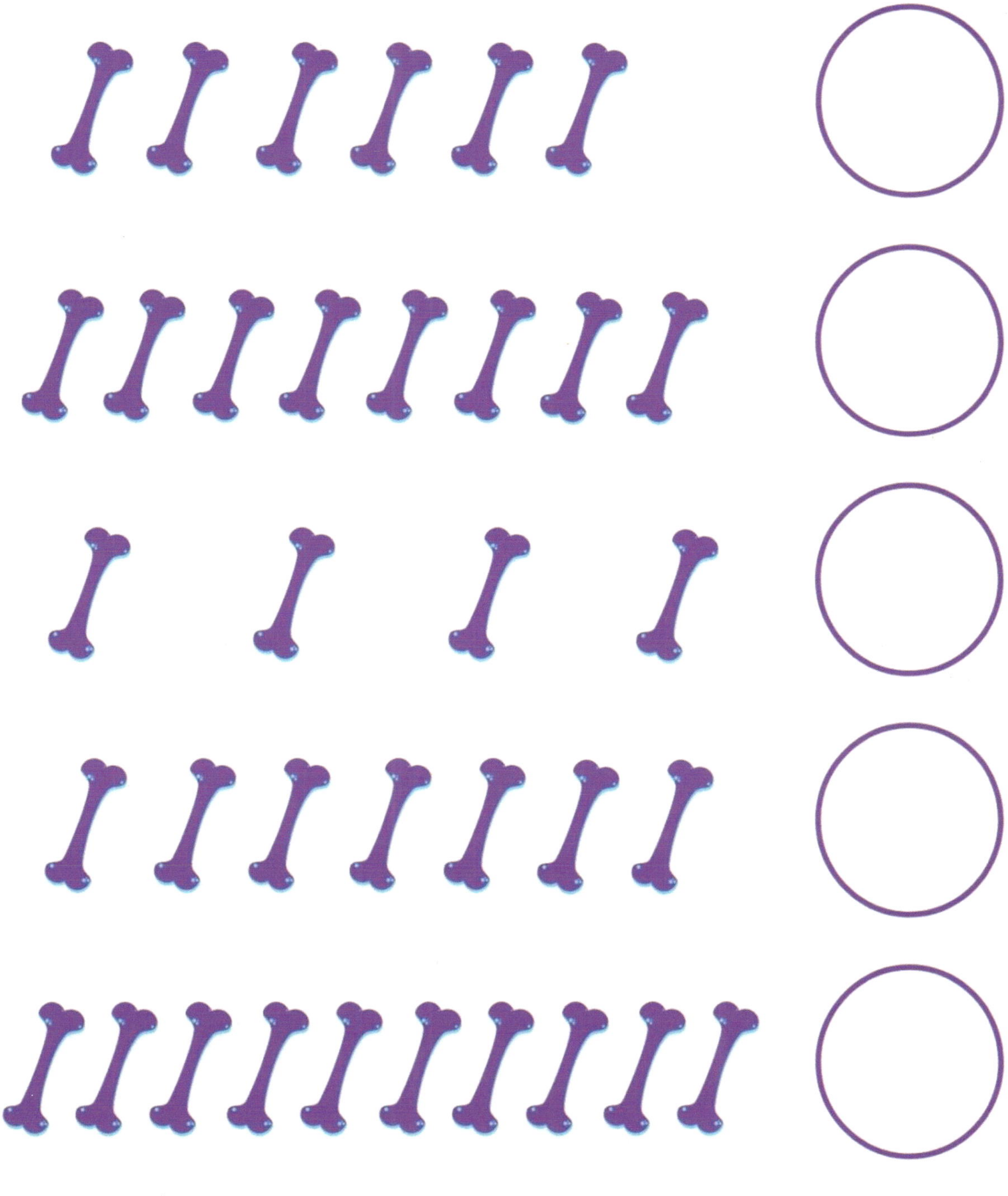

SIMPLE SUBTRACTION

Write subtraction sentences to match the pictures in each box.

□ − □ = □

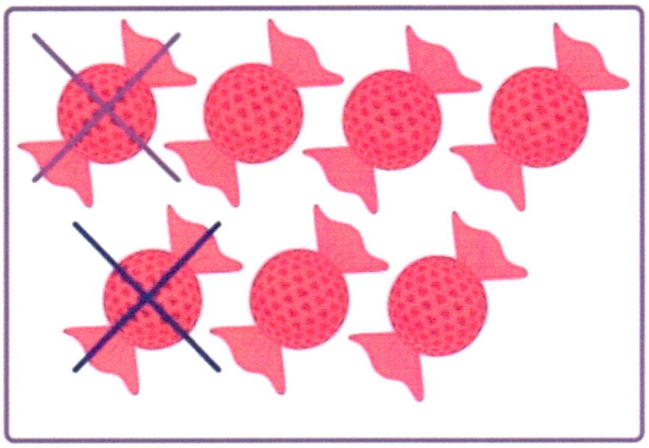

□ − □ = □

□ − □ = □

COLOR-BY-NUMBER

Level: Stretch Myself

Answers:

10 = Pink

8 = Yellow

6 = Purple

9 = Orange

5 = Brown

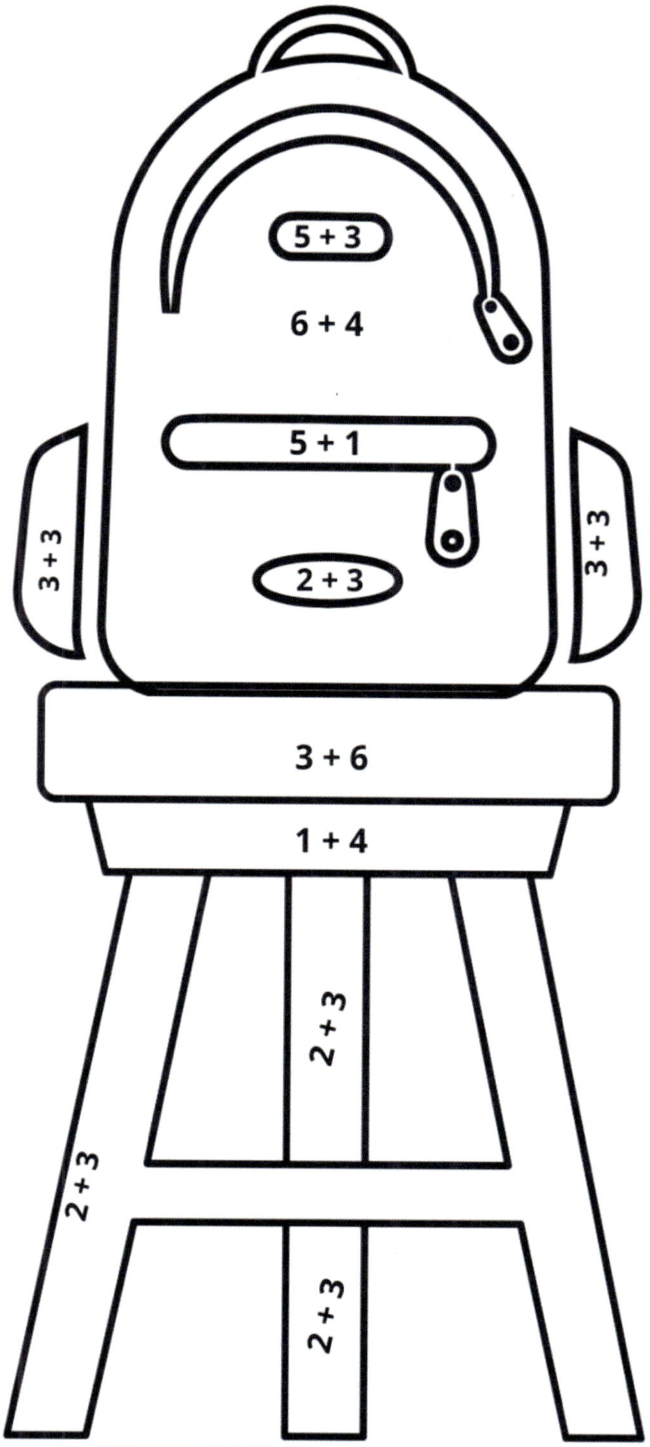

ADDITION

Add the numbers shown on the hands and write the answer in the space provided.

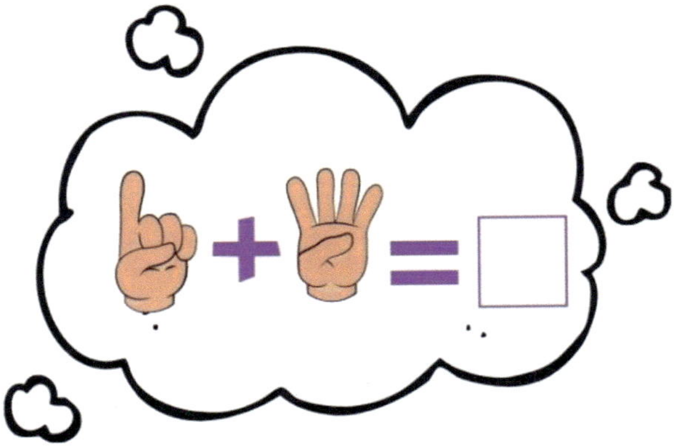

SUBTRACTION

Subtract the numbers shown on the hands and write the answer in the space provided.

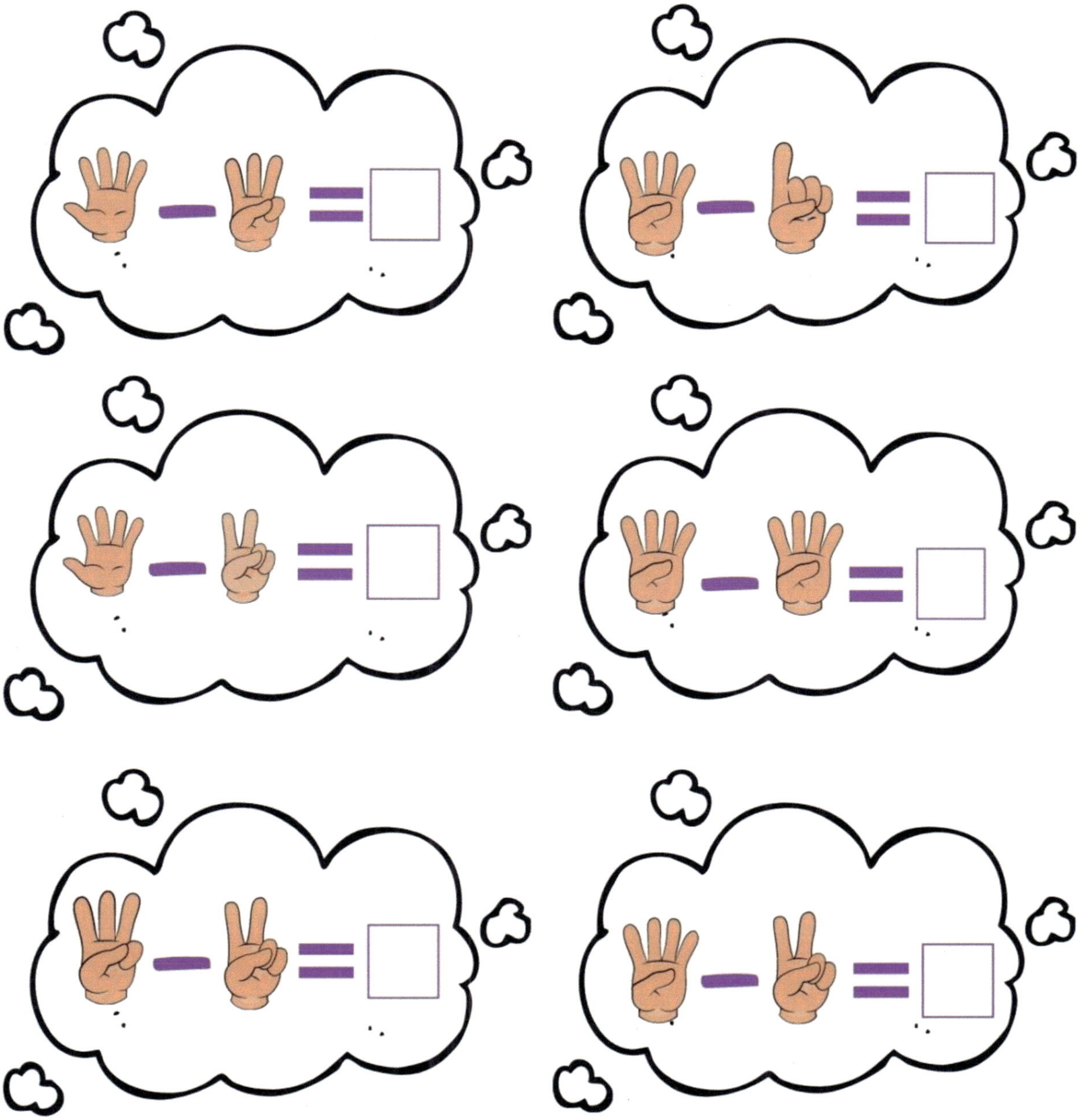

MATH DICE GAME

Add up the dice pictured. Put the answer in the space provided.

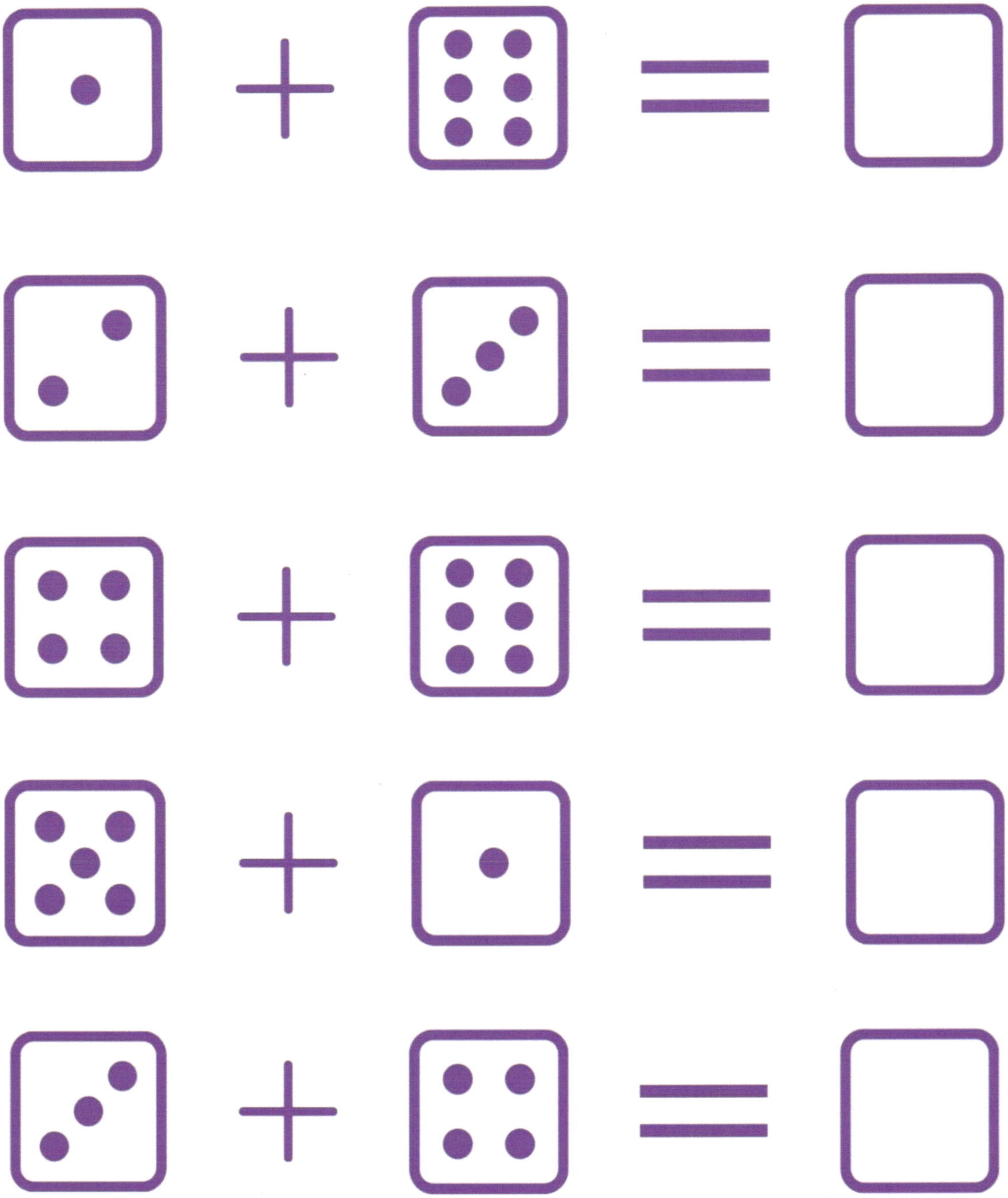

SINGLE DIGIT ADDITION

Let's add the numbers!

1 + 2 =	3 + 4 =	2 + 2 =	6 + 3 =
7 + 1 =	3 + 7 =	9 + 1 =	5 + 3 =
6 + 4 =	2 + 5 =	5 + 5 =	7 + 2 =
5 + 1 =	1 + 8 =	4 + 2 =	4 + 4 =

Single Digit Subtraction

Let's subtract the numbers!

6 -3	7 -5	10 -3	8 -2
8 -7	6 -1	3 -2	5 -2
7 -3	10 -5	8 -1	7 -7
5 -3	9 -4	8 -3	9 -3
8 -4	6 -2	8 -8	7 -6

FILL IN THE MISSING DOTS

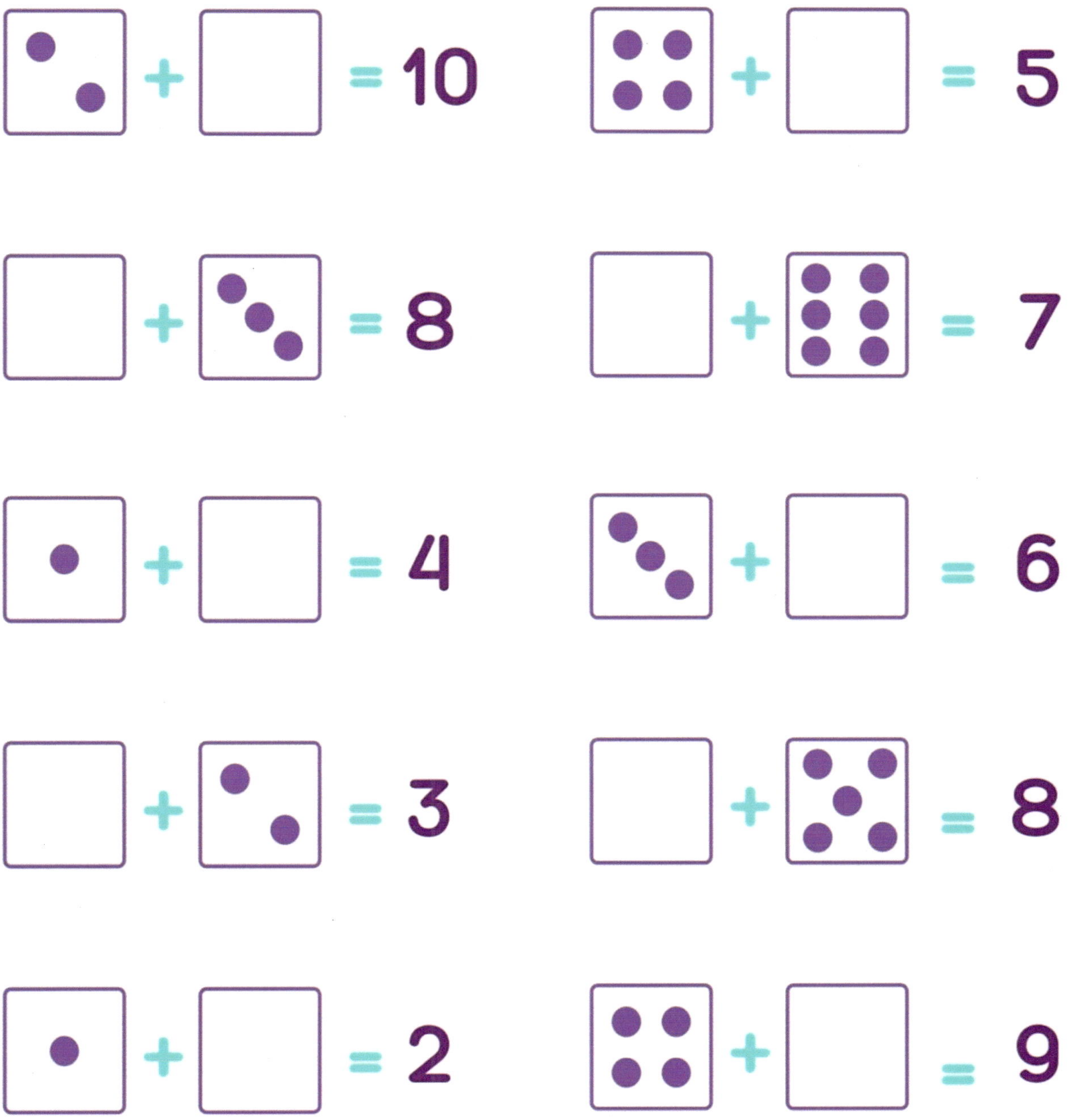

CLOCK TIME BY THE HOUR!

Write the time that is on each clock.

MATCH THE CLOCK

Draw a line to connect the matching time.

7:00

3:00

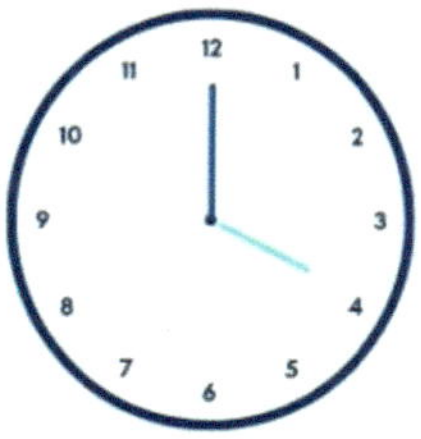

4:00

10:00

MAKE YOUR OWN CLOCK!

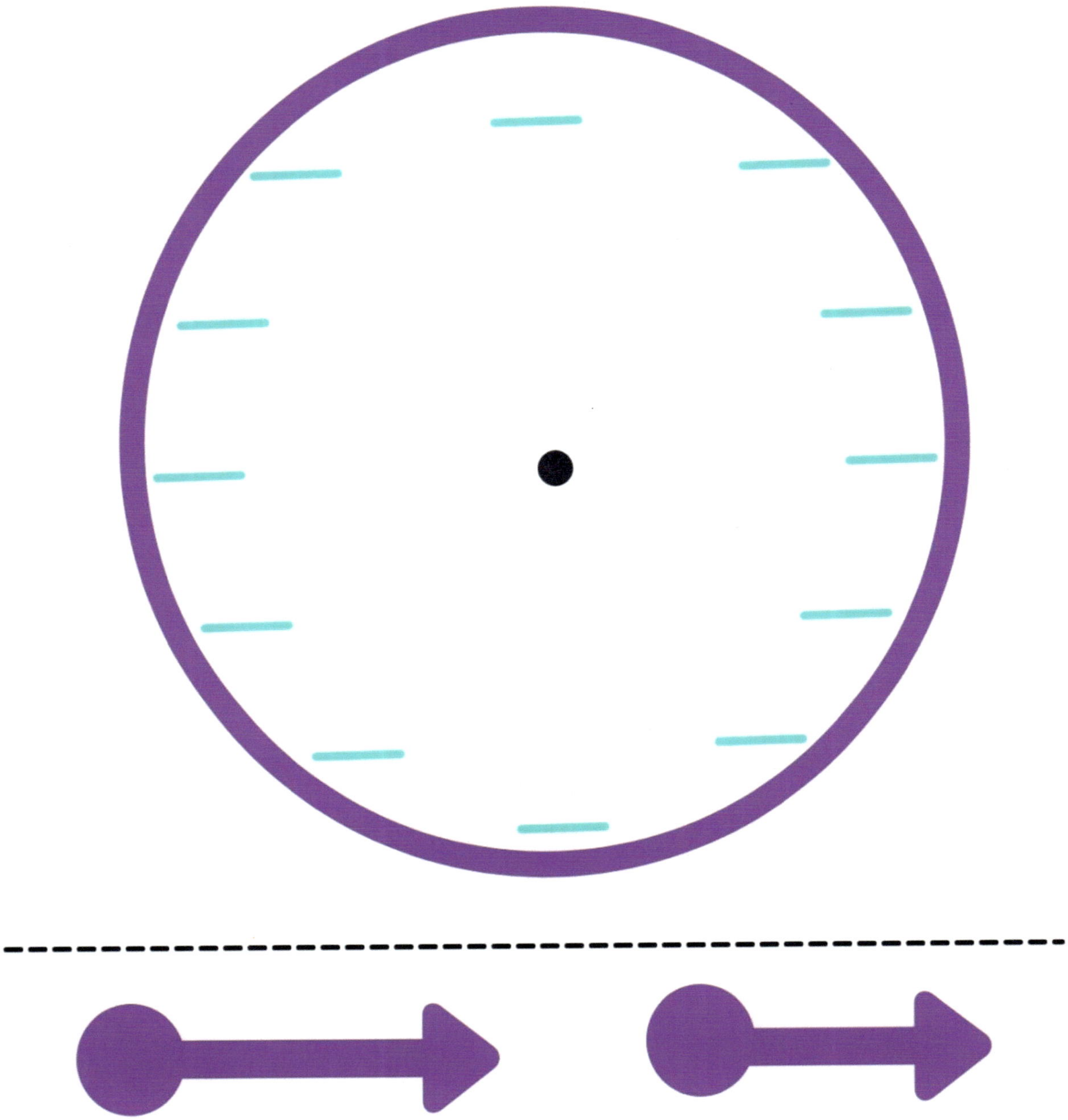

ADDITION AND SUBTRACTION

$$\begin{array}{r} 7 \\ +\ 3 \\ \hline \end{array} \qquad \begin{array}{r} 5 \\ +\ 1 \\ \hline \end{array} \qquad \begin{array}{r} 3 \\ +\ 6 \\ \hline \end{array}$$

$$\begin{array}{r} 6 \\ +\ 2 \\ \hline \end{array} \qquad \begin{array}{r} 4 \\ +\ 4 \\ \hline \end{array} \qquad \begin{array}{r} 5 \\ +\ 2 \\ \hline \end{array}$$

---

6 - 3 = ☐ 7 - 7 = ☐

8 - 2 = ☐ 8 - 3 = ☐

5 - 1 = ☐ 8 - 6 = ☐

7 - 5 = ☐ 4 - 2 = ☐

TRACING SHAPES

Trace the shapes of different sizes

1. **Squares**

2. **Triangles**

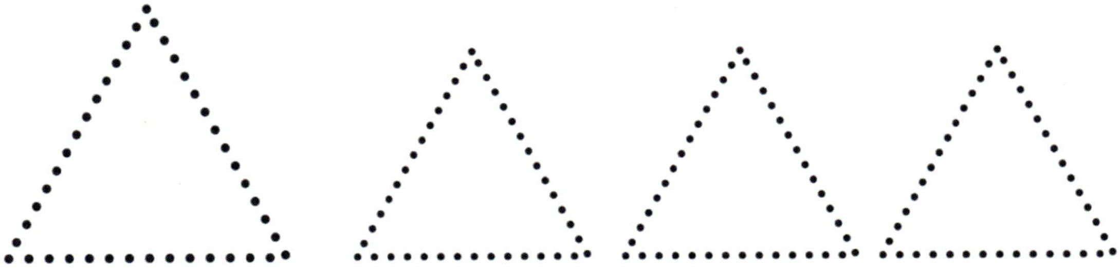

3. **Circles**

4. **Rectangles**

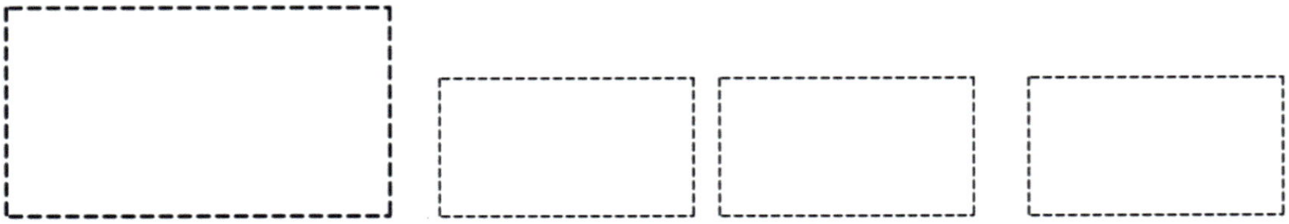

MATCH THE SHAPES

Draw a line from the object on the left to the shape that matches it.

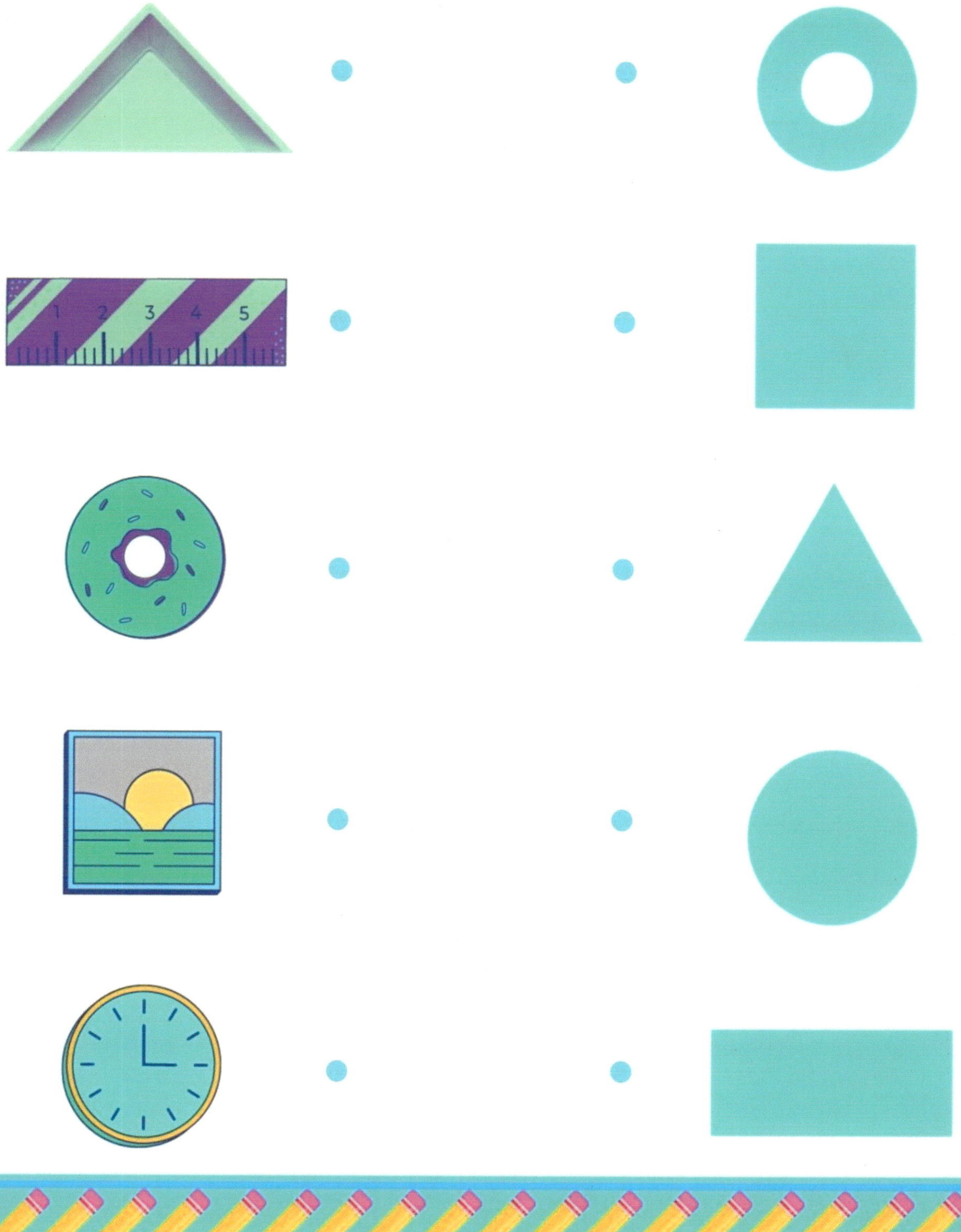

UNDERSTANDING SHAPES

LET'S DRAW AND COLOR!

I. COLOR THE SQUARES BLUE, THE CIRCLES RED, AND THE TRIANGLES GREEN.

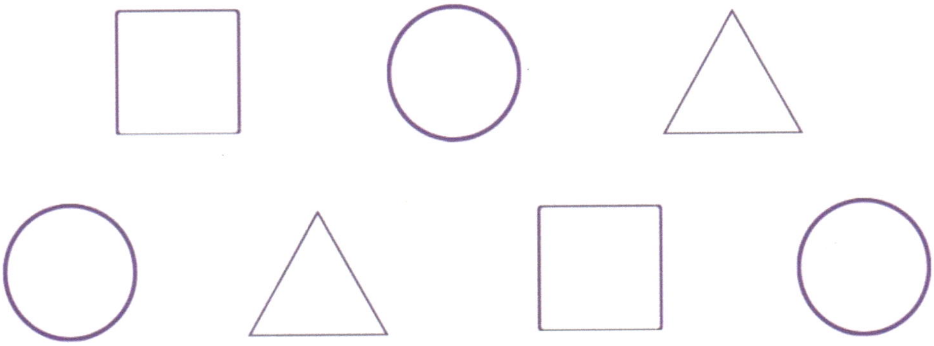

II. DRAW YOUR FAVORITE TWO SHAPES IN THE SPACE PROVIDED BELOW.

TRACE EACH SHAPE

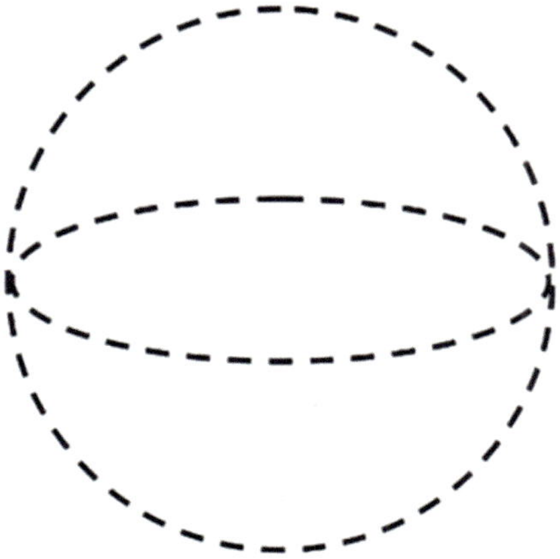

Sphere

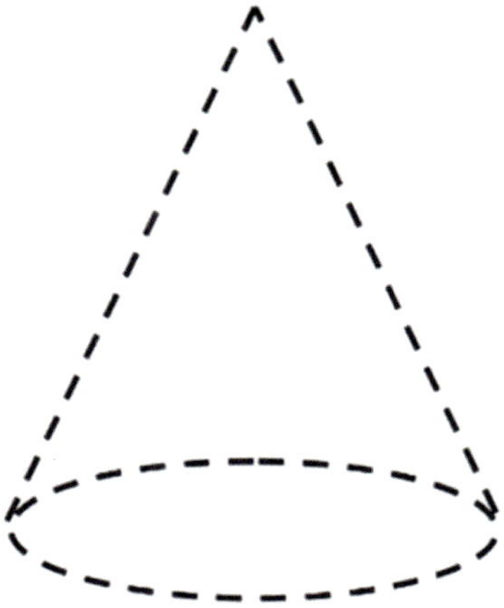

Cone

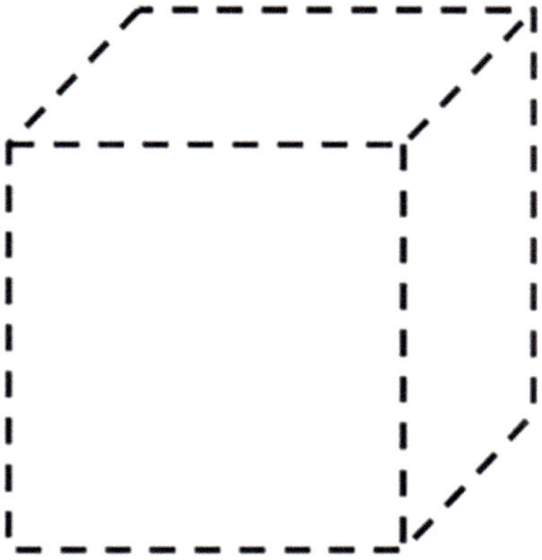

Cube

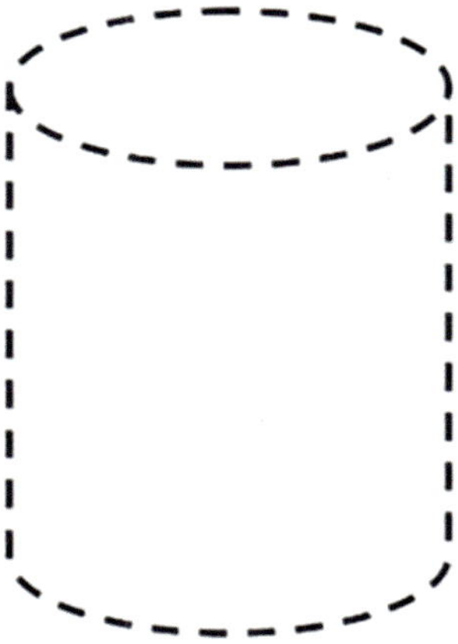

Cylinder

Made in the USA
Las Vegas, NV
07 November 2026

57672506R00019